Agrivoltaics

A Sustainable Integration
of Solar Energy and Agriculture

Giuseppe Saturno

Translation from Italian

ISBN: 9798398185706

To all those who believe in innovation and sustainability.

This dedication is for you, pioneers of change and custodians of the Earth. In the world in which we live, where clean energy and sustainable agriculture have become imperative, you are the ones blazing new trails.

CONTENTS

Giuseppe Saturno is a passionate expert in Permaculture and renewable energy, with a fervent dedication to a fair and better world.

With over 15 years of experience in the field of Permaculture, he has developed in-depth skills and knowledge in promoting sustainable agricultural systems and ecological design practices. From a young age, he has shown a strong interest in the environment and sustainability.

After completing his studies in Permaculture, he dedicated his life to spreading awareness about the benefits of an integrated approach to environmental design and agriculture. His training has provided him with a solid theoretical and practical foundation for creating sustainable solutions that improve people's lives and conserve natural resources.

In addition to his specialization in Permaculture, he has become an expert in the field of renewable energy. His passion for preserving the environment has driven him to explore and adopt sustainable energy solutions. Through his experience, he has gained in-depth expertise in the design, installation, and management of solar photovoltaic systems, wind energy systems, and other clean energy technologies.

What sets him apart is his vision, perhaps utopian, of an equitable world. He is an enthusiastic dreamer who firmly believes that every individual can make a difference in building a more sustainable society. His dedication and passion have driven him to share his knowledge and inspire others to take concrete actions to preserve the environment and create a better future for all. In addition to his practical work, Giuseppe is also a popular speaker. In fact, he has given several lectures and workshops on topics related to Permaculture, renewable energy, and sustainability in general, spreading his vision and encouraging people to take all possible actions to preserve the environment.

INTRODUCTION

Definition of Agrivoltaics

Agrivoltaics (also Agrovoltaics) is a concept that combines agriculture with photovoltaic energy into a single integrated infrastructure. It is a practice of using the same agricultural land to grow plants or trees while simultaneously installing solar photovoltaic panels to generate electricity.

In a nutshell, Agrivoltaics represents the synergistic integration of agriculture and solar energy on a single site, where solar panels are installed on top of cultivated fields or on specific structures such as pergolas or greenhouses.

This combination offers several advantages. First, the shading provided by solar panels can reduce sunlight intensity and ambient temperature, creating a favorable microclimate environment for crops, especially in areas with high insolation.

Shading can also reduce water evaporation from the soil, helping to conserve water resources.

Agrivoltaics also enables dual land use, optimizing the

use of agricultural space without compromising food production or energy efficiency. Combined food and energy production can diversify farmers' income and thus provide additional income.

Overall, Agrivoltaics aims to achieve a balance between sustainable agriculture and renewable energy production, promoting ecological resilience and resource efficiency.

Importance of solar energy and sustainable agriculture

Solar energy and sustainable agriculture are both critically important for the future of our planet. Let's see why:

1. Solar Energy:

- **Renewability**: Solar energy is a virtually inexhaustible source of energy unlike fossil energy sources. By harnessing solar energy, we can reduce dependence on nonrenewable sources and mitigate the negative effects of fossil fuels on the environment, such as air pollution and the emission of greenhouse gases responsible for climate change.

- **Global accessibility**: The sun is a resource available everywhere in the world, although in varying amounts depending on region. Harnessing this energy allows us to generate electricity in a decentralized way, bringing clean energy even to rural or remote communities that may have difficulty accessing the traditional power grid.

- **Reducing carbon emissions**: Solar power generation, unlike coal- or gas-fired power plants, does not emit CO_2 or other greenhouse gases during operation. A statement that is now trite, yes, but it serves to clarify definitively the concepts we will go on to see later.

2. Sustainable Agriculture:

- **Food security**: Truly sustainable agriculture adopts practices that maintain soil fertility, conserve water and reduce the use of synthetic pesticides and fertilizers. This helps preserve the long-term productivity of agricultural lands, ensuring food security for future generations. And not only for future ones...

- **Resource conservation**: This type of agriculture aims to make efficient use of natural resources such as water or human labor, reducing waste and minimizing environmental impact. This includes adopting efficient irrigation practices, managing wastewater, and using renewable energy sources for necessary agricultural operations.

- **Biodiversity and ecosystem health**: Sustainable agriculture obviously promotes biodiversity by diversifying crops, conserving natural habitats and reducing the use of harmful chemicals. These measures help to preserve ecosystems, maintain ecological balance and protect the health of plants, animals and humans.

- **Climate Change Resilience**: Agriculture done sustainably adopts only practices that improve the resilience of crops and agricultural ecosystems to climate

8

change. This includes the selection of resistant varieties, the use of soil conservation techniques, and the sustainable management of water resources. In this way, it contributes not only to mitigating the negative impacts of climate change on agriculture itself, but also on society as a whole.

Therefore, we can already say that solar energy and agriculture are the two key pillars for the future. Integrating energy production into agriculture can lead to more efficient and ecologically sustainable food production, improving the resilience of farming communities to climate change.

Benefits and Challenges of Agrivoltaics

Agrivoltaics offers a number of significant benefits, but also challenges that have yet to be addressed. Let us explore further what we have hinted at:

Benefits:

Efficient land use: Agrivoltaics allows the same land to be used for solar energy production and agriculture and/or animal farming. This dual land use maximizes the efficiency and yield of the agricultural area, avoiding the need to dedicate separate land for agriculture and solar energy.

Crop-friendly microclimate: The installation of solar panels provides shading to the crops below, reducing sunlight intensity and ambient temperature. This creates a cooler and more humid microclimate, which can promote plant growth, especially in regions with high temperatures or water scarcity.

Water savings: Photovoltaic systems can reduce water evaporation from the soil, as the shading provided by the panels reduces direct exposure to the sun. This helps conserve water resources, making irrigation more efficient and reducing the amount of water needed to grow plants.

Diversification of income sources: Agrivoltaics offers the opportunity to generate additional income for farmers. In addition to the production of food or other crops, the sale of the solar energy produced can provide a stable source of income. This diversification of income sources can make farms more resilient and economically sustainable.

Reducing carbon emissions: Obviously, the use of solar energy in the agricultural context contributes to the transition to a low-carbon economy.

...And Challenges:

System Design and Planning: The design and installation of an Agrivoltaics system requires careful planning. Various factors need to be considered, such as orientation of solar panels, height of support structures, and choice of compatible crops. Careful planning is essential to maximize the benefits of both agriculture and solar energy.

Land use competition: Agrivoltaics clearly requires adequate space for panel installation. This can result in land use competition between agriculture and solar energy production. A balance needs to be struck between the two activities and the impact on agriculture and food production needs to be carefully evaluated. (And it is precisely in Agrivoltaics that there can be this balance).

Crop management and maintenance: Crop management in an Agrivoltaic system is still a challenge. It is important to consider shading and access to sunlight for the plants below as well as space to work with them. In addition, maintenance of the solar panels must also be considered to ensure proper system operation.

Costs: The installation of an Agrivoltaic system can require a significant initial investment. Costs include the purchase of solar panels, support facilities, and installation of the system. However, the long-term benefits, such as reduced energy costs and diversification of agricultural income, offset these initial costs. And it is also true that one can start with a minimal and scalable installation and then increase it later.

Integration and regulation: Integration of Agrivoltaics into the context of policies and regulations can be a catnip. Clear regulations and appropriate incentives still need to be developed to promote the adoption of Agrivoltaics. In addition, collaborations among different stakeholders, including farmers, energy companies, and governments, may be needed to facilitate the implementation of large-scale Agrivoltaic projects.

Addressing these challenges would require at the government level careful planning, collaboration across sectors, and a long-term commitment to develop innovative solutions. Despite these challenges and some hurdles to overcome, Agrivoltaics offers significant potential for bringing agriculture and energy production together.

It should also be added that **Agrivoltaics is a very new technique, there are not many globally valid studies or even long-time experts!**

We still have to experiment quite a lot and that is why every experience made is like an extra brick.

With this book we continue precisely to set the basics and try to give everyone the tools to start and make their own experience.

BASICS OF SOLAR ENERGY

Basic concepts on Photovoltaics

As everyone knows, solar energy is the energy that comes from sunlight, and it is renewable and free, which means that it does not deplete and contribute to the depletion of such valuable natural resources. This energy can be converted into usable energy through various technologies, such as photovoltaic panels or solar thermal panels.

Photovoltaic panels convert sunlight into electricity using photovoltaic cells. When photons of light strike the photovoltaic cells, they generate a stream of electrons that creates an electric current. This is the current used to power everything we need or to be stored in batteries.

Solar thermal panels, on the other hand, absorb heat from the sun to heat water or other fluids. This heat can be used for domestic purposes such as space heating, domestic hot water production or greenhouse heating.

Solar systems can be installed on canopies, building roofs, land or other surfaces exposed to the sun. As already mentioned, depending on where you are located, things change.

To get an idea of the accessibility and intensity in your area I recommend searching for "PHOTOVOLTAIC GEOGRAPHICAL INFORMATION SYSTEM" on Google.

To read a little more about it:

A photovoltaic panel, also called a solar panel, is a device that uses the photovoltaic effect to convert sunlight into electricity. Its operation is based on scientific principles related to semi-conduction and the photovoltaic effect.

Inside a photovoltaic panel, there are photovoltaic cells consisting of semiconductor materials, usually silicon. These materials are treated in such a way as to create a p-n layer, that is, a layer with an electron-rich zone (n) and a zone with a surplus of gaps (p). This configuration creates a p-n junction, which is critical to the operation of photovoltaic cells.

When photons of sunlight strike the photovoltaic cells, they are absorbed by the semiconductor materials. The energy from the photons is transmitted to the electrons in the electron-rich zone (n) and excites them, allowing them to overcome the energy barrier present in the p-n junction. This phenomenon creates a charge separation within the cell, with the electrons moving outward along the electrical circuit connected to the photovoltaic panel.

The flow of electrons creates an electric current used precisely to power electrical devices or be stored in batteries for future use. The photovoltaic panel is capable of producing electricity as long as it is exposed to sunlight, even not necessarily direct sunlight, and its ability to generate excited electrons depends on the intensity and frequency of the incident photons. To ensure constant current flow and appropriate voltage, photovoltaic panels are often connected in series or parallel to form modules or arrays. These modules can be *combined into larger solar systems to meet the energy needs of a building or facility. It is important to note that the efficiency of solar panels can vary depending on the technology used, the materials used, and environmental conditions.*

Ongoing developments in research and technological innovation aim to improve the efficiency and performance of photovoltaic panels, making them increasingly competitive as a sustainable and environmentally friendly source of energy.

Common units of measurement for solar panels include watt (W) and kilowatt (kW), which represent the rated power of the solar panel. The power rating indicates the amount of energy the panel can generate under certain standard solar irradiance conditions.

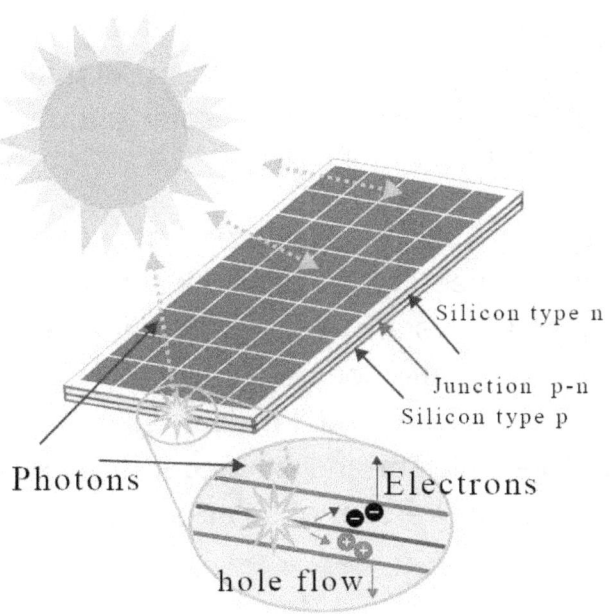

Silicon type n

Junction p-n
Silicon type p

Photons

Electrons

hole flow

Storage of the energy produced by solar panels and its transformation into alternating current (AC) are two important aspects of the complete solar system.

Energy storage: When solar panels generate electricity, it can be used immediately or can be stored for future use. Energy storage is especially important when the energy generated exceeds immediate demand or when the sun is not present, such as at night or during cloudy days. Batteries are often used for solar energy storage. Excess electricity produced by solar panels is stored in batteries for later use when needed.

Conversion to alternating current (AC): The energy produced by solar panels is direct current (DC), but most appliances and power grids use alternating current (AC). Therefore, it is necessary to convert the DC energy produced by the solar panels into usable AC energy. This conversion is done using a solar **inverter**. The inverter converts DC energy into AC energy, making it compatible with appliances and allowing the solar energy to be fed into the household or public grid.

The efficiency of energy storage and conversion to AC is an important aspect to consider when designing and installing a solar system. Energy storage technologies, such as batteries, are constantly improving in capacity, efficiency, and durability. Likewise, solar inverters are becoming more efficient and advanced to ensure reliable and high-quality conversion of solar energy to AC.

Solar technologies used in Agrivoltaics

Here are some of the technologies used:

Photovoltaic solar panels: Photovoltaic solar panels are the most widely used solar technology in Agrivoltaics. They can be installed on elevated supports or on structures such as pergolas, sheds or canopies within the agricultural area, simultaneously allowing the plants below to grow.

Agricultural solar roofs: This technology takes advantage of spaces above sheds or agricultural structures for the installation of solar photovoltaic panels. Solar agricultural roofs not only generate solar energy, but also provide protection from the elements to the structures below, such as agricultural warehouses and equipment.

Solar shade structures: These structures are designed to provide shade for agricultural crops, protecting them from excessive direct sunlight, while at the same time generating solar energy. They can be made of transparent or semi-transparent photovoltaic solar panels that allow light to pass through, providing adequate amounts of sunlight to plants. We will look at the different types in a moment.

Solar pumps: Solar pumps are used for irrigating agricultural crops and obviously use solar energy. These pumps are powered directly by panels, eliminating the need for an external electrical connection or a diesel generator. Solar pumps are useful in rural or remote areas where access to traditional electricity is limited.

Solar lighting: Solar lighting is often used in agricultural spaces to illuminate work areas, buildings or paths at night. Solar lighting systems harness solar energy to power LED lights, offering a sustainable and economical alternative to traditional lighting.

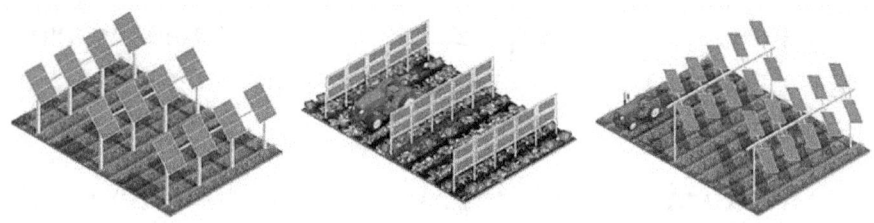

Variables for the design of an Agrivoltaic system include the choice of structure (fixed or mobile), height above ground, materials and characteristics, module spacing, angle of inclination, and the type and percentage of shading desired.

An Agrivoltaic system consists of an operating system (fixed or tracking), a supporting structure, and a ground anchorage. All types of solar modules can be used, but the most common are those with silicon solar cells, which account for most of the global photovoltaic market. These modules consist of a sheet of glass on the front and a white covering film on the back, mounted on a metal frame. The solar cells are connected in series and laminated between the two elements. A metal frame is used for mounting and stability.

The Agrivoltaic system can be fixed (vertical, horizontal, tilted) or variable (single- or dual-axis tracking). In tracking systems, the modules follow the movement of the sun using a tracking mechanism. Single-axis tracking follows the sun horizontally, while biaxial tracking optimizes both elevation and azimuth. This type of system can maximize energy output, but involves higher purchase and maintenance costs.

The support structure must be adapted to the needs of the system, taking into account the clear height and row spacing. Good clear height ensures an even distribution of light under the system and allows access for agricultural machinery. Ground anchorage or foundation is important to ensure the stability of the Agrivoltaic system. In addition to permanent concrete solutions, there are environmentally friendly alternatives such as pile foundations or the Spinnanker (or Spinanchor) system, which can be removed without leaving a trace.

BASIC PRINCIPLES OF SUSTAINABLE AGRICULTURE

Fundamental Concepts

Sustainable agriculture is an approach that aims to grow food and produce agricultural resources in an ecologically sound, socially equitable and economical way. This approach takes into account the needs of present generations without compromising the ability of future generations to meet their own needs. Here are the main points:

Conservation of natural resources: Sustainable agriculture is committed to conserving natural resources, such as soil, water and biodiversity. This means adopting soil management practices that reduce erosion, compaction and nutrient depletion. In addition, efficient irrigation methods are promoted to conserve water and natural habitats are preserved to promote biodiversity and zero pollution.

Reduced use of chemical inputs: Sustainable agriculture seeks to reduce the use of chemical fertilizers and synthetic pesticides that can have negative effects on the environment, human health, and soil and water quality. The use of alternative agricultural practices, such as crop rotation, intercropping, composting, and biological pest and disease control, is promoted.

Promotion of biodiversity and ecosystems: Sustainable agriculture is concerned with promoting biodiversity and healthy agricultural ecosystems. This includes conservation of native species, creation of habitats for wildlife, promotion of natural pollination, and integrated pest and disease management.

Water conservation: This type of agriculture adopts only water management practices that reduce water use and waste. This may include drip irrigation, rainwater harvesting and use, watershed management, and protection of water resources from excessive pollution.

Valuing local communities and farm workers: Agriculture done right is also committed to creating decent working conditions and promoting the participation of local communities in decisions about agricultural practices. This includes respecting the rights of farm workers, promoting local employment, and developing local food systems that promote food security and community resilience.

Sustainable agriculture is a holistic approach that integrates environmental, social and economic principles to create a balanced agricultural system that is sustainable over the long term. It promotes environmental health, human health and economic prosperity by seeking to achieve a balance between the needs of agricultural production and the conservation of natural resources.

Importance of natural resource conservation

Resource conservation is of critical importance to the well-being of our planet and ourselves. **And we should all realize this by now...**

Natural resources, such as soil, water, forests and the famous biodiversity, are the very essence of life on Earth, and their balance is crucial to the maintenance of ecosystems and our survival.

Conservation of natural resources is important for several reasons that we often forget. First, ecological balance depends on the availability and proper use of these resources. Fertile soil is essential for food production and plant growth. Without proper soil management, agriculture becomes less productive and food security is jeopardized.

In addition, water is a vital resource for life. Its availability and quality are critical for maintaining aquatic ecosystems and also for meeting the needs of human communities. Forest conservation is crucial for several reasons. Forests are habitats for numerous plant and animal species, contributing to biodiversity.

They also play a key role in carbon sequestration and climate change mitigation. Deforestation and unsustainable use of forests can lead to loss of biodiversity, soil erosion and increased greenhouse gas emissions, contributing to the acceleration of climate change.

Let's repeat, biodiversity conservation is of utmost importance to preserve the diversity of life on Earth!

It alone provides essential ecosystem services, such as plant pollination, climate regulation, water purification and protection from natural disasters. In addition, many plant and animal species are very useful sources of food, medicine and natural materials for humans.

Ultimately, the conservation of natural resources is critical to ensuring ecological balance, species survival, food security, and the well-being of human communities. Promoting sustainable natural resource management and preservation practices is a collective endeavor that requires the active participation of governments, institutions, businesses and, above all, individuals. Only through resource conservation can we secure our future.

"When they have polluted the last river, cut down the last tree, caught the last bison, caught the last fish, only then will they realize that they cannot eat the money accumulated in their banks".

Reducing the environmental impact of conventional agriculture

Reducing the environmental impact of conventional agriculture refers to the practice of adopting measures and strategies to mitigate the negative effects that conventional agriculture can have on the environment. Conventional agriculture, which often makes extensive use of chemical fertilizers, synthetic pesticides, and intensive production methods, has a number of harmful environmental impacts, such as soil and water pollution, loss of biodiversity, and emission of greenhouse gases. To reduce the negative environmental impact of conventional agriculture, several strategies and practices have already been developed:

Soil management practices: Practices that improve soil quality and reduce erosion are promoted, such as **crop rotation**, use of **vegetative cover crops**, conservation tillage and **composting**. These practices help preserve soil structure, maintain soil fertility, and reduce the risk of erosion.

Integrated pest and disease management: Integrated pest and disease management approaches that reduce reliance on synthetic pesticides are encouraged. This includes the use of biological and natural methods of pest and disease control, selection of resistant varieties, and crop rotation.

Reducing the use of chemical inputs: Efforts are made to limit the use of chemical fertilizers and synthetic pesticides by trying more natural alternatives. This may include the adoption of targeted fertilization techniques, the use of organic fertilizers, such as composting and animal manure.

Water conservation: Water management practices that reduce the use and waste of water in agricultural activities are promoted. This may include drip irrigation and mulching, the use of water-efficient irrigation systems, and also, where possible, choosing crops based on water needs.

Promotion of agricultural biodiversity: Efforts are made to conserve and the promotion of agricultural biodiversity, such as by growing only local plants, creating and preserving habitats for wildlife, and protecting insects such as bees. Agricultural biodiversity is what most promotes the resilience of agricultural ecosystems and contributes to crop stability.

...**And Permaculture**, to which I prefer to devote a separate paragraph for a very brief and very reductive introduction...

The book that will bring you closer to Permaculture permanently.

Permaculture

"Though the problems of the world are increasingly complex, the solutions remain embarrassingly simple"

Bill Mollison

Why, who and when

Giuseppe Saturno

Permaculture

Permaculture is a system of design and practice that is based on the principles of ecology and ethics to create sustainable systems that meet the needs of people and nature. The term "Permaculture" comes from the combination of the words **"permanent agriculture"** and "**culture**," highlighting the goal of creating agricultural and social systems that are sustainable over the long term.

It was developed in the 1970s in Australia by **Bill Mollison** and **David Holmgren**. The two founders combined their knowledge of ecology, agriculture, anthropology and systems design to develop an integrated and holistic approach to sustainable systems design. In 1978, Mollison and Holmgren published the book "*Permaculture One*," which is considered the foundational text of Permaculture.

Permaculture is based on three basic ethics: **caring for the Earth, caring for people**, and **sharing resources equitably**. These ethics guide permaculturists' decisions and actions in designing and managing systems. Permaculture also incorporates twelve design principles that provide guidelines for creating sustainable systems, including using resources sparingly, designing for resilience, and promoting diversity.

Permaculture goes beyond agriculture and includes a broader vision of sustainable living systems. It applies not only to agriculture, but also to architecture, landscape design, water management, renewable energy, education, economics, and community. The goal is to create integrated systems that are in harmony with natural processes, promote biodiversity, are energy efficient, and meet people's needs in a sustainable way.

It has had a significant impact in the environmental movement and sustainable design and has become a philosophy of life for many who seek to live more harmoniously with the natural environment. Permaculture has been adopted worldwide, with permaculture projects and communities implementing permaculture principles and practices to create sustainable and resilient living patterns.

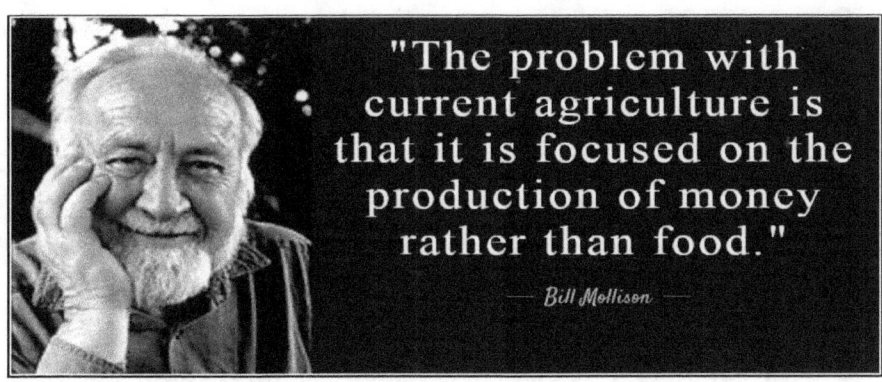

"The problem with current agriculture is that it is focused on the production of money rather than food."

— Bill Mollison —

DESIGN AND PLANNING OF AN AGRIVOLTAIC SYSTEM.

Choosing the ideal location and orientation

Choosing the ideal location and orientation for an Agrivoltaic installation depends on several factors that need to be considered and requires a thorough analysis of local conditions, agricultural crop requirements, and technical and economic capabilities. Here are some key points to consider when selecting a location:

Sun exposure: It is essential to place the Agrivoltaic plant in an area with good sun exposure. This means that the area should be directly exposed to sunlight for most of the day. An accurate analysis of solar irradiance can be helpful in determining the suitability of a specific area. In Chapter 2 I pointed you to a useful site for an accurate analysis.

Shading: It is important to assess the presence of any obstacles that could cause significant shading on the Agrivoltaic system. Trees, buildings, or other structures can reduce the efficiency of the PV system and affect plant growth.

Terrain topography: The topography of the terrain can affect the efficiency of the Agrivoltaic system. It is preferable to choose relatively flat terrain to facilitate the installation of solar panels and ensure adequate sun exposure.

Soil conditions and drainage: It is important to consider the quality of the soil and its drainage capacity. Well-drained soil helps prevent water accumulation and the risk of stagnation, which could damage both solar panels and agricultural crops.

Accessibility and infrastructure: It is necessary to assess the accessibility of the area to facilitate the installation, maintenance, and operation of the Agrivoltaic system. In addition, it is important to consider the presence of infrastructure, such as access to electricity and the distribution grid, to ensure proper connection of the PV system.

Agronomic considerations: It is critical to consider the specific needs of the agricultural crops that will be grown in the Agrivoltaic area. Some crops may require different exposure conditions or benefit from certain soil characteristics. The interaction between solar energy and plant cultivation must be carefully evaluated to maximize the benefits to both systems.

Selection of crops compatible with Agrivoltaics

The selection of crops compatible with Agrivoltaics is an important aspect of ensuring the success and productivity of the system. The goal is to find crops that can coexist symbiotically with solar panels, making the best use of space and optimizing available resources. Here are some factors to consider when selecting crops:

Crop height and bearing: Choose crops that do not obstruct solar radiation on solar panels. Crops with low height or vertical habit, such as herbs, leafy vegetables, flowers or compact shrubs, are often more suitable as they do not interfere with solar energy production.

Growth cycle: It is important to select crops with growth cycles compatible with solar energy production. For example, annual or biennial crops that are harvested or replaced before solar panel leaves become shaded may be an appropriate choice.

Shade tolerance: Although efforts are made to reduce shading, it is inevitable that solar panels will create shade on crops below. Therefore, it is important to select crops that are tolerant to partial shade and can continue to grow and develop under these conditions.

Water needs: Consider the water needs of the crops and the availability of water resources in the area. Choosing crops that require similar amounts of water can facilitate irrigation management in the Agrivoltaic system.

Biodiversity and ecological synergies: Promoting biodiversity in the Agrivoltaic system can have significant ecological benefits. Choosing crops that attract pollinating insects, repel pests, or promote biological control of pests can help create an ecological balance in the system.

Economic choice: Although this would not be viewed well by Sitting Bull and the founders of Permaculture, one should evaluate the profitability of the selected crops in relation to production costs and sales markets. Choosing commercially valuable crops that are in demand in the market can contribute to the economic sustainability of the Agrivoltaic system.

The ultimate goal is to create a combination of crops and solar panels that support each other and maximize sustainable food and energy production.

Within Agrivoltaics, there are several crops that are well suited to this combined solar energy production and agriculture system. These crops are selected based on their characteristics and ability to grow and thrive in an environment that includes solar panels. Let's look at some examples of crops that are often considered suitable for Agrivoltaics.

Herbs, such as **mint**, **parsley**, **basil**, **sage** and **lavender**, are a popular choice. They are low height plants that do not require prolonged direct sunlight and can be easily grown between solar panels.

Leafy greens, such as **lettuce**, **spinach** and **arugula**, are also suitable for Agrivoltaics. These crops are characterized by a rapid growth cycle.

Low-growing climbing plants, such as **climbing peas, beans or zucchini**, can grow vertically without interfering with solar panels, making effective use of available space.

Some **low-elevation flowers**, such as **dwarf sunflowers o r marigolds**, can be grown in the Agrivoltaic environment, adding aesthetic value to the area and attractiveness to beneficial pollinators.

In addition, some varieties of **fruiting shrubs**, such as **blueberries**, **blackberries** or **strawberries**, are suitable for Agrivoltaic. These compact shrubs can be grown between solar panels without creating significant shade.

IMPACT OF SHADING ON CROPS

Study of the effects of shading

Studies of the effects of shading on plants are critical to understanding how the presence of structures such as solar panels in agriculture can affect crop growth and health. These studies allow us to assess the positive or negative impacts of shading on plants and to adopt appropriate strategies to maximize productivity in the Agrivoltaic environment.

When discussing plant shading, it is important to consider several aspects. First, the intensity and duration of shading varies depending on the location of the solar panels, the angle, the size of the structures, and the pattern of the sun during the day.

The effects of shading on plants depend on various factors, including the type of crop, the period of shading, the intensity of reduced sunlight, and the surrounding environmental conditions. In general, shading can affect the following:

Photosynthesis and plant growth: Shading reduces the intensity of sunlight reaching plants, affecting photosynthesis, the process by which plants convert sunlight into chemical energy for growth. Reduced exposure to sunlight can limit plants' ability to produce nutrients and grow optimally.

Morphological development: Shading can affect the morphological behavior of plants, such as by resulting in increased height growth (positive phototropism) to reach

32

sunlight or reduced lateral branching.

Flower and fruit production: Shading may affect flower and fruit production. Some plants may have reduced flowering ability or a decrease in the quality and quantity of fruits produced due to reduced exposure to sunlight.

Competition with weeds: Shading can also affect plants' competition with weeds. Reduced sunlight can promote the growth of weeds, which compete with agricultural crops for water, nutrients and space.

To fully understand the effects of shading on plants in the Agrivoltaic environment, it is necessary to conduct specific studies on different crops, assessing their tolerance to shade and adapting agronomic practices accordingly. Several strategies can be adopted to mitigate the negative effects of shading, such as choosing crops adapted to partial shade, optimizing solar panel placement, and implementing soil and irrigation management techniques.

Studies on the effects of shading on plants in Agrivoltaics are an evolving field of research, as the goal is to find an optimal balance between solar energy production and agricultural efficiency. This research allows us to adopt increasingly sustainable agrivoltaic practices and maximize the benefits in terms of both renewable energy production and agricultural production. As mentioned above, however, there are still no rules that apply to everyone, so it is important to arm yourself with all the knowledge already available and experiment for yourself!

Protection from sun damage and extreme weather events. Shade reduces evaporation and maintains soil moisture. Decreases soil temperature on hot days.

Adaptation of crops to shade

However, agricultural crops can adapt to shade in various ways to optimize their growth and production. These are some of the adaptation mechanisms that plants use to cope with shade:

Positive phototropism: Many plants exhibit a response called positive phototropism, which means they tend to grow toward the light. When shaded, plants extend their stems or leaves toward available light sources to maximize the absorption of solar energy.

Increased photosynthetic efficiency: Shade plants can also adapt by increasing the efficiency of the photosynthesis process. This can be done by modifying leaf architecture, such as by developing thinner or wider leaves to capture more light, or by increasing the concentration of chlorophyll in leaves to maximize the uptake of available light.

Reduced lateral growth: Shaded plants may reduce lateral growth, focusing instead on vertical growth to reach available light. This can lead to increased plant height and reduced lateral branching.

Adaptation of flowering times: Some plants can even adjust flowering times. They may flower at times when there is greater availability of sunlight or they may extend flowering duration to maximize seed or fruit production.

Development of shade tolerance mechanisms: Some plants are able to develop shade tolerance mechanisms, such as an increased ability to withstand conditions of reduced sunlight. These plants can adapt to shady conditions and maintain acceptable growth and production despite reduced amounts of light.

It is important to note that the ability of crops to adapt to shade may vary depending on the species and the specific conditions of the environment. Some may be more adaptable than others and may be preferred in agri-voltaic systems where shading is a more significant factor (higher panel density).

Maximizing agricultural production efficiency in Agrivoltaics

To maximize the efficiency of agricultural production in Agrivoltaics, several strategies can be adopted. Some key considerations are listed below:

Crop choice: Mundane but always underestimated, choose crops suitable for the Agrivoltaic environment, considering exposure, shading, and water requirements. Opting for short-cycle crops or perennial crops adapted to partial shading can allow for higher yields.

Crop rotation: Implementing crop rotation, another aforementioned old and forgotten technique, helps preserve soil fertility, reduce the risk of disease and pests, and maximize resource efficiency. Altering crops sequentially over different sections of the agri-voltaic area promotes balanced use of all resources.

Complementary crops: Integrate crops that complement each other and promote ecological synergies. For example, some aromatic plants can repel insects harmful to other crops or attract beneficial pollinators. Choosing positively interacting crops can promote a more resilient and efficient Agrivoltaic system. There are very clear, proven tables that apply everywhere.

Irrigation management: Closely monitor crop water requirements and adopt an appropriate irrigation system to ensure optimal water supply. The use of technologies such as soil moisture sensors and drip irrigation systems can enable more precise and targeted water management. There are very simple and made-in-Italy systems such as "Arduino" that allow with just a few euros to have the situation always under control. "Arduino Grow Station" is another term to Google.

Weed control: Maintain effective weed control to avoid competition with agricultural crops. The use of mechanical methods, such as plowing or green manure, along with soil cover techniques such as mulching, have always helped reduce weed growth.

Integrated disease and pest monitoring and management: Pay attention to timely monitoring of plant diseases and pests by adopting integrated management strategies that include cultural and biological methods (including chemical but targeted methods if necessary). Preventive and careful management can minimize production losses and pesticide use.

Agrivoltaic system performance monitoring: Constantly evaluate the efficiency of the Agrivoltaic system by collecting your own data on crop yield, solar energy production, and resource utilization. This information provides invaluable insights for making improvements and optimizing overall system performance.

IRRIGATION AND WATER RESOURCES MANAGEMENT.

Efficient and sustainable irrigation systems

Irrigation systems are critically important for ensuring responsible use of water resources and maximizing agricultural production. These systems are designed to optimize water use efficiency, reduce waste, and minimize negative impacts on the environment. Let's look at some of the main irrigation systems used to achieve these goals.

One of the most popular systems is **drip irrigation**, which involves delivering water directly to plant roots through small drippers or porous pipes. This system reduces water losses due to evaporation and soil erosion, allowing more targeted and efficient use of available water resources. Another effective system is **micro-aspersion** irrigation, which uses small jets of water to irrigate plants. This system allows even distribution of water over the soil, reducing losses and allowing precise control over the amount of water delivered to plants. Another option is **sub-irrigation**, which involves partially or completely submerging the soil to allow plant roots to absorb the necessary water. This system is particularly suitable for soils with good water retention capacity.

Spray irrigation is another common technique, using sprinklers to distribute water over crops evenly. It is important to use high-quality sprayers to reduce losses due to evaporation and drift effect.

In addition, modern technology has introduced precision irrigation systems, which combine the use of soil moisture sensors and automatic controls to deliver water exactly when and where it is needed. This avoids over- or under-irrigation, reducing waste and optimizing overall water efficiency.

Another important consideration is the use of solar energy to power irrigation systems. This approach harnesses the sun's renewable energy to run pumps and irrigation systems, reducing the environmental impact associated with the use of fossil fuels.

To maximize the efficiency and sustainability of irrigation systems, it is also essential to adopt soil management practices, such as covering the soil with organic material (the aforementioned mulching). In addition, closely monitoring crop water needs and adapting the irrigation regime according to climatic conditions can help reduce waste and improve water management.

The adoption of efficient and sustainable irrigation systems certainly contributes to reducing water consumption, preserving soil quality, limiting erosion and groundwater pollution, and promoting responsible water resource management in the agricultural sector. These systems are an important step toward a much more sustainable and efficient agriculture.

Rainwater harvesting and utilization

Rainwater harvesting and use play a key role in wise and sustainable water resource management. Here are some of the main insights:

Water resource conservation: Water is the most precious resource and is at the same time limited. Collecting and using rainwater allows us to reduce our dependence on traditional freshwater sources such as rivers, lakes and aquifers. This helps to preserve water resources available for essential uses and reduce pressure on water supply. We will have quite a few problems in the future if we do not also start managing rainwater well.

Reducing water stress: In many regions of the world, water scarcity is now a quixotic problem. Rainwater harvesting and use can provide an additional source of water for non-potable purposes such as crop irrigation, animal washing, surface cleaning, and cooling. This helps to reduce water stress and ensure sustainability of water uses.

Reducing water pollution: Rainwater harvesting can help reduce the load of pollutants that end up in surface water and groundwater. Rainwater can carry pollutants such as fertilizers, pesticides, oils, and sediment from soil and urban surfaces. By collecting and treating these waters, pollution of water resources can be prevented or reduced.

Flood reduction: Rainwater harvesting can help reduce the risk of localized flooding. By collecting rainwater through drainage systems and reservoirs, direct flow to streams and drainage channels can be restricted, preventing overloading of drainage systems and reducing the risk of flooding. Little thing, but it all adds up, even that the straw that breaks the camel's back....

Economic savings: Using rainwater undoubtedly reduces the costs associated with traditional water supply. Installing rainwater harvesting systems may require an initial investment, but in the long run it can lead to significant savings in water costs.

Promoting sustainability: Rainwater harvesting and use are sustainable practices that promote responsible water management and environmental conservation. To reiterate, these are the practices that help conserve and protect water resources.

In conclusion, rainwater harvesting and use are important to ensure water conservation, reduce water stress, prevent water pollution, manage flooding, and promote sustainability. These practices are an effective way to responsibly and efficiently use a valuable natural resource such as rainwater.

Integrated water management for agriculture and energy

Integrated water management for agriculture and energy is an approach that aims to coordinate and optimize the use of water resources to simultaneously meet the needs of agriculture and energy production. This approach recognizes the interconnection between water use for agricultural and energy purposes and seeks to address the challenges and opportunities that arise from this interaction.

In the context of agriculture, water is essential for crop irrigation and food production. However, the demand for water for agricultural purposes can be significant and put a strain on available water resources. On the other hand, energy production also requires a considerable amount of water, such as for cooling thermal power plants or for biofuel production. This is not exactly the topic of this book, but two words must be spent.

Integrated water management for agriculture and energy seeks to address the challenges associated with these two activities, seeking to maximize water use efficiency and minimize negative impacts on the environment. This approach is based on a number of strategies and practices, including:

Planning and coordination: This involves planning water use for both agriculture and energy in an integrated way, considering local needs, water availability, and priorities. Cooperation between the two sectors and collaboration between stakeholders are key to the effective management of water resources.

Efficient water use: Adopting practices that improve water use efficiency is a key aspect of integrated management. This includes the use of efficient irrigation systems, optimized irrigation scheduling (crop needs), soil moisture monitoring, and implementation of precision irrigation techniques.

Use of sustainable energy sources: Promoting the use of renewable energy sources reduces the environmental impact associated with energy production. Adopting solar, wind or hydroelectric technologies helps reduce dependence on water-intensive sources.

Wastewater management: Water from agricultural

production can be treated and reused for irrigation purposes or hydropower generation. Wastewater recycling is an important strategy for optimizing the use of water resources.

Monitoring and evaluation: It is important to monitor water use and evaluate the effectiveness of implemented management strategies. Monitoring water resources, consumption, and environmental impacts allows for possible corrections or improvements to the management system.

MONITORING AND CONTROL OF ENVIRONMENTAL PARAMETERS

Importance of monitoring environmental parameters

Monitoring of environmental parameters is essential for assessing the impact of human activities on the environment and for being able to take effective mitigation and conservation measures. Through monitoring, the right decisions can be made to preserve the health of ecosystems and promote environmental sustainability.

Monitoring of environmental parameters involves the systematic and regular collection of data and information regarding various aspects of the environment, such as air quality, water quality, biodiversity, pollution, climate and other environmental indicators. This process includes installing survey instruments and conducting sampling, analysis and observations to assess the state of the environment and any changes over time. Environmental monitoring thus provides the basis for that crucial information that allows problems to be identified, the effectiveness of policies and management practices to be evaluated, and the right decisions to be made.

Monitoring Technologies in Agrivoltaics

Here are some examples of technologies:

Solar radiation sensors: These sensors measure the intensity and direction of solar radiation incident on the agricultural area. This information makes it possible to assess the efficiency of solar panels in absorbing solar energy and to identify any areas of shading that may affect energy production.

Soil moisture sensors: These sensors measure soil moisture content at different depths. This allows soil water conditions to be monitored and irrigation practices to be optimized, **avoiding waste and shortages.**

Meteorological sensors: These sensors measure various meteorological parameters such as air temperature, relative humidity, and wind speed and direction. This information is important for understanding the growing environment of crops and the effect of climate also on the efficiency of the solar panels themselves. **At temperatures that are too high or too low, their output changes!**

Crop monitoring systems: These systems use sensors and advanced technologies to monitor crop growth, soil quality, and other agronomic parameters. For example, they can measure plant height, plant cover, leaf chlorophyll and other characteristics to assess crop health and photosynthetic efficiency.

Energy monitoring: These systems measure energy production from solar panels and monitor the efficiency of energy conversion and transformation systems. This makes it possible to evaluate the performance of the systems and identify any problems or failures.

The combined use of these monitoring technologies provides a comprehensive view of the interactions between solar energy, agricultural crops, and the surrounding environment. This allows farmers and operators of agrivoltaic systems to customize and thus improve crop management, irrigation, energy management, and overall system optimization.

As mentioned earlier, Agrivoltaics is not and cannot yet be an exact science given the complexity and number of factors involved. It is precisely personal and local experience that will make the difference in the long run. While therefore having a solid foundation on both PV and agriculture, the fact remains that every location on this planet will be different.

ECONOMIC BENEFITS OF AGRIVOLTAICS

Reducing energy costs for agriculture

Reducing energy costs in agriculture is a key goal for improving the efficiency and sustainability of all agricultural operations. There are several strategies that can be adopted to achieve this goal.

First is the obvious use of renewable energy. The installation of solar panels and the use of wind turbines provide a clean, low-cost source of energy to power all agricultural operations. In addition, taking measures to improve energy efficiency plays a crucial role in reducing costs. This can include the use of advanced technologies, such as high-efficiency electric motors, energy-efficient **LED lighting**, and thermal insulation in agricultural facilities. These are the practices that help to optimize energy use more and reduce waste. Another important aspect is the optimization of irrigation systems. Irrigation represents one of the agricultural activities that requires a considerable amount of energy. Using efficient irrigation systems, such as drip irrigation or precision irrigation, can reduce the associated energy consumption as well as improve water use efficiency.

The use of advanced technologies, such as **smart sensors and automation systems**, offers additional opportunities to optimize energy use in agriculture. These technologies allow irrigation, lighting and other activities to be precisely monitored and regulated, reducing waste.

Finally, **cooperation between farms** can help reduce

energy costs. Cooperative networking enables energy exchange between farms, allowing energy resources to be shared and overall costs to be reduced. Cooperating and sharing is always better. I also recommend that you learn about **"energy communities."**

Reducing energy costs as usual not only leads to economic savings for farmers, but is also an important contribution to environmental sustainability.

Opportunities for additional income through energy production

Energy production represents a significant opportunity to generate additional income for farmers and ranchers. There are several ways through which this opportunity can be exploited:

Sale of energy: Farmers can install renewable energy production systems, such as solar panels or wind turbines, and sell the energy produced to the electricity grid. This is a practice that is now common throughout Europe and allows them to earn money from the energy generated, based on purchase agreements and government incentives.

Self-consumption of energy: Farmers can use the energy produced to meet their own energy needs within their farming operations. This allows them to reduce the amount of purchased power and achieve significant savings not only in the long term.

Diversification of activities: The installation of renewable energy production facilities is also a form of diversification of agricultural activities. Both farmers and ranchers can combine energy production with traditional farming activities, creating new income opportunities and reducing dependence on a single source of income.

Incentives and subsidies: In many countries, there are government programs and incentives aimed at promoting energy production from renewable sources. Anyone can benefit from such programs, which offer subsidies, affordable tariffs, tax breaks and even grant funding for the installation and operation of energy production facilities.

Integrating energy production into agricultural activities thus offers new economic opportunities in every way.

Economic evaluation of Agrivoltaics

Agrivoltaics, by combining agriculture and solar energy production, offers a number of attractive economic opportunities. The economic evaluation of such a system is critical to understand whether it is a financially beneficial option. First, **initial investment costs** must be considered. Setting up an Agrivoltaic system obviously requires the purchase and installation of facilities, implementation of a suitable irrigation system, and other infrastructure requirements. Costs vary depending on the size of the system, the technology used, and the specifications of the land. Keep in mind that sometimes it is better to start small without investing large sums. This allows you to practice and begin to discover the typical characteristics of the area and their influence on the complete system that can then be expanded.

Once the system, whether large or small, is built, the solar energy is finally converted into electricity that can be used on the farm to meet its own needs. Alternatively, it will be possible to sell it to the national power grid, earning income through payments from the operator.

Another aspect to consider is the possible **diversification and expansion of agricultural activities**. Agrivoltaics allows farmers to combine energy production with new traditional agricultural activities, creating new sources of income. For example, they can grow plants suitable for solar panel shading that were not possible to grow previously. Given the fact that the panels change the microclimate anyway, it is clear that this offers other possibilities to be explored. However, the economic evaluation of Agrivoltaics must also take into account possible risks and uncertainties. Fluctuations in energy prices, changes in climatic conditions, and regulatory changes can affect the profitability of the project. It is important to conduct an analysis, considering all the long-term costs and benefits, in order to assess whether Agrivoltaics represents a profitable economic opportunity.

Incentives

Agrivoltaics is a technology that precisely requires fairly expensive facilities, costing up to 30-40% more than a traditional ground-mounted PV system. These expenses represent a significant burden that agricultural entrepreneurs often cannot bear on their own. To enable the development of this technology, the use of economic incentives becomes crucial. So far, the spread of agrivoltaic systems has been hindered by a lack of regulatory inclusion in the incentive system.

CASE STUDIES AND SUCCESSES IN AGRIVOLTAICS

Designing an Agrivoltaic Plant

Here are some examples of successful agrivoltaic projects that have been implemented in different parts of the world:

"SolarCrop" in Japan: This project implemented solar panels suspended above rice growing fields. The shading provided by the solar panels helped reduce heat stress on rice plants and improve crop yields. The project demonstrated that Agrivoltaics can promote food production and renewable energy in a limited area of land.

"Ciel et Terre" in France: This project used floating solar panels on reservoirs to generate solar power. The floating solar panels were placed on a reservoir and provided electricity for the local power grid. Using the reservoirs to install the solar panels maximized land efficiency and conserved water resources.

"Food and Energy Training and Education" (FEED) in the United States: This project created an Agrivoltaic model combining food production and renewable energy. Solar panels were installed on elevated structures to create shade and provide an environment conducive to growing vegetables with high nutritional value. The project demonstrated that Agrivoltaics can contribute to sustainable food production and clean energy generation.

"AgriPV" in the Netherlands: This project combined

agriculture and solar energy by installing solar panels on agricultural greenhouses. The solar panels provided energy for lighting and irrigation of the greenhouses, thus reducing energy costs and environmental impact. The project demonstrated that Agrivoltaics can improve energy efficiency in agriculture and enable higher crop production.

In Italy, here are some examples:

"Tarquinia": Enel Green Power has started construction of the largest Agrivoltaic Solar Park in Italy, located in Tarquinia, in the province of Viterbo. The plant will have a capacity of about 170 MW and will be able to produce an average of 280 GWh of renewable energy per year. In addition to contributing significantly to clean energy production, the solar farm will avoid the emission of about 130,000 tons of CO_2 per year and replace the consumption of 26 million cubic meters of fossil gas. Double-sided photovoltaic module technology mounted on solar trackers will be used to maximize energy efficiency. In addition, the solar farm will be integrated with agricultural activities, growing forage, borage and olive trees in the vacant areas between the panels and in the buffer strips of overhead power lines. The project is an important step toward sustainable energy production and land enhancement.

"Turning Point": In Puglia, Italy, the first Agrivoltaic plant in the country and one of the first in Europe is born. The story is told by Nicola Mele, an entrepreneur who focuses on organic farming, research and a new 8 MW plant. The link between Agrivoltaic and sustainability can be explained in many ways, but to make it concrete there is nothing better than a practical example provided by the story of a farm born in Puglia, which was the first in Italy (and among the first in Europe and perhaps the world) to have the foresight to create an Agrivoltaic plant in 2011. Today, the entrepreneur who created the conditions for the birth of that plant, which had a capacity of almost 1 MW, has an even more ambitious project: to build an 8 MW one, reconciling energy production from renewable sources with agriculture. In this specific case, the intention is to start wine production according to organic criteria, convinced that linking PV with agriculture is beneficial.

But that's not all: underlying the project is the belief that reconciling the two worlds is beneficial for agriculture. This belief is supported by scientific evidence from studies conducted by Maurizio Boselli, former professor of viticulture at the University of Verona, and Giuseppe Ferrara, professor of arboriculture and fruit growing at the University of Bari. They both share a common history with Nicola Mele, the entrepreneur who was instrumental in the establishment of the Apulian company Svolta, where the agrophotovoltaic plant was built, and the subsequent creation of "I Prodotti della Svolta". This company is among the founding members of AIAS, the Italian Association for Sustainable Agrivoltaics.

The story of Svolta began in the Veneto region of Italy in 2008, when the University of Verona decided to conduct research to understand the potential of reconciling photovoltaic production with agriculture. A pergola structure is created to support photovoltaic panels, using agricultural and viticultural materials and adopting techniques used in the construction of Trentino and Verona pergolas, avoiding the use of agricultural foundations. The goal is to understand what benefits can be obtained by growing vegetables and vines shaded by photovoltaic panels.

In the same year, thanks to the collaboration between the team of researchers of Arboriculture at the University of Bari, coordinated by Professor Giuseppe Ferrara, and Maurizio Boselli, former professor of Viticulture at the University of Verona, research is started on the feasibility of an Agrivoltaic system in Apulia, using wine vineyards and taking advantage of the region's peculiar climatic characteristics.

The objective of this research was to study and highlight opportunities to introduce a photovoltaic system to improve grape conditions. Due to climate change and rising temperatures, wine grapes ripen early without having time to develop aromas.

This is where Nicola Mele, an IT entrepreneur with experience at the Olivetti research center and a successful track record in information technology, comes in. The Roggero family, which is involved in the Svolta farm, calls him to start a cutting-edge agricultural and energy enterprise in Puglia. The company "Svolta" (an acronym for Solare VOLTaico Ambiente-Agricoltura) is established, where the first agrivoltaic installations are made and research on vineyards shaded by photovoltaic panels is started in 2009, in collaboration with Professor Boselli at the University of Verona.

On the farm located in Laterza, in an area close to Gioia

del Colle, Santeramo and Matera, various experimental research on wine cultivation is conducted. On a total area of 7 hectares, an Agrivoltaic system of 972 kW is installed on 4 hectares, with the panels placed more than two meters high. It is cultivated both inside and outside the Agrivoltaic area, comparing the results. Starting in 2019, in collaboration with the Basile Caramia Institute of Locorotondo, which conducted analysis and vinification of the agrivoltaic grapes, it was possible to see the effectiveness of the project: the wines produced have rich and intense aromatic characteristics.

"Mazara del Vallo" in Sicily: Engie has inaugurated Italy's largest Agrivoltaic Park in Mazara del Vallo, Sicily. The plant covers 115 hectares and has a capacity of 66 MW, and is part of a Corporate PPA (Power Purchase Agreement) contractual model between Engie and Amazon. This is the first Agrivoltaic park built in Italy and the first based on this type of agreement between private companies. The construction of the plant was made possible thanks to a €100 million green loan financed by Cdp, Societé Générale and BNP Paribas. In addition to producing clean energy, the goal of the Agrivoltaic Park is to cultivate fields with plants such as vines, olive trees, almond trees, and aromatic and medicinal plants.

In addition, a second 38 MW Agrivoltaic site is planned to be built in Paternò, in the province of Catania, as part of the agreement between Engie and Amazon. In total, the two plants will have an installed capacity of 104 MW and the energy produced will be used to power Amazon's operations in Italy.

The Agrivoltaic park in Mazara del Vallo uses state-of-the-art technology, with double-sided solar panels mounted on single-axis trackers that capture both direct light and light reflected from the surrounding land, optimizing energy production. This configuration makes it possible to reduce the area required for the photovoltaic system and maximize agricultural effectiveness.

During the construction of the Mazara del Vallo plant, 150 people were employed.

Positive impact of Agrivoltaics on farming communities.

Local job creation: The development and implementation of agrivoltaic projects can generate new local jobs. Specialized skills such as installers, electricians, and solar technicians are needed during the installation phase of solar panels and the construction of support facilities. These jobs can be done by members of the community itself, providing local employment opportunities and contributing to the economic growth of the region.

In addition, once the Agrivoltaic system is operational, ongoing maintenance and management activities are required. This includes cleaning solar panels, maintaining irrigation systems, and monitoring energy efficiency. These tasks can be performed by local workers, creating long-term stable employment in farming communities.

Land enhancement: The implementation of agrivoltaic projects can contribute to the enhancement of agricultural and rural land. The integration of solar technologies with traditional agricultural activities creates a modern and sustainable image of agriculture, promoting the attractiveness of the area for investment and tourism.

The visual appearance of an Agrivoltaic system, with solar panels integrated into crops or over fields, can give a distinctive character to the agricultural landscape. This can generate interest from visitors and tourists who wish to learn about and experience innovative and sustainable agricultural models.

In addition, the adoption of Agrivoltaics can promote better agricultural land management. The efficient use of agricultural space through the integration of agricultural activities and solar energy production can contribute to resource conservation and environmental protection. This sustainable approach to agriculture can foster the creation of agri-tourism networks, promoting the direct sale of agricultural products and the enhancement of local traditions.

Overall, local job creation and land enhancement are two significant benefits of Agrivoltaics on farming communities. These factors not only contribute to the local economy, but also strengthen rural identity, promoting sustainable development and attractiveness of rural areas.

Lessons learned and best practices in the implementation of Agrivoltaics

During the implementation of Agrivoltaics, important lessons were learned that can guide the process effectively and sustainably.

Summarizing them:

One of the most significant lessons is the selection of suitable crops. It is essential to select crops that can thrive under the shade of solar panels, such as low-elevation plants or varieties that require less direct sunlight. Careful design and proper engineering are also crucial to ensure the reliability and safety of the Agrivoltaic system over the long term, taking into consideration soil conditions, local regulations, and durable materials.

It is important to plan **irrigation** according to crop needs and reduce water waste through the use of drip or low-consumption systems. Rainwater harvesting and use can also contribute to the water sustainability of agriculture.

Regular **monitoring and maintenance** of the Agrivoltaic system is essential to ensure maximum energy and agricultural performance. This includes monitoring solar panel efficiency, evaluating irrigation, and checking crop conditions. Regular cleaning of solar panels is especially important to ensure that there is no significant reduction in efficiency due to dirt or dust accumulation.

Finally, **stakeholder involvement** is critical to the success of Agrivoltaics. Farmers, solar energy experts, local authorities, and surrounding communities need to be involved early in the project. Collaboration and knowledge sharing foster better understanding and adoption of Agrivoltaics. In addition, it is important to tailor solutions to the specific needs of farming communities, promoting the integration of Agrivoltaics into their farming practices.

TECHNICAL AND REGULATORY CHALLENGES TO FACE

Agrivoltaic implementation presents several technical and regulatory challenges that must be addressed to ensure its success.

Infrastructure integration: The installation of solar systems within agricultural areas requires proper infrastructure integration. Interconnection with the existing power grid must be considered to ensure a stable and secure energy flow. In addition, the design and installation of support structures for solar panels must be well planned to minimize the impact on agricultural activities.

Water resource management: The efficient use of water is an increasingly important challenge in Agrivoltaics and beyond. It is necessary to balance the irrigation needs of agricultural crops with the water consumption required by solar panels. Water resource management must be optimized to avoid waste and ensure equitable distribution of water among crops.

Optimizing energy efficiency: Energy efficiency is a key factor in the success of any project. It is necessary to maximize solar energy production by choosing efficient photovoltaic technologies and optimizing the orientation and tilt of solar panels. At the same time, it is important to reduce energy losses during transmission and conversion.

Regulation and regulations: Agrivoltaic implementation requires adherence to a number of regulations and standards, sometimes unclear or absent. These may relate to installation and grid connection, safety issues, and environmental regulations. It would be important for regulatory aspects to be clear and well defined to facilitate Agrivoltaic adoption and ensure compliance with applicable laws. Unfortunately, it is not up to us...

Awareness and acceptance: Agrivoltaics is a relatively new practice that requires greater awareness and acceptance by stakeholders. There is a need to inform farmers, local communities, and authorities about the potential and benefits of Agrivoltaics. This may involve outreach efforts, training, and active stakeholder involvement to overcome any resistance and encourage adoption of this sustainable practice.

Addressing these challenges requires effective collaboration between farmers, solar energy experts, authorities, and local communities. An integrated approach that considers both technical and regulatory aspects is needed to ensure a successful transition to Agrivoltaics as a sustainable practice in agriculture in the near future.

Innovations and future developments in Agrivoltaics

Agrivoltaics is, as mentioned above, a constantly evolving field that offers many opportunities for future innovation and development. There are several areas in which significant progress is expected to be seen:

One of the main areas of innovation involves photovoltaic technologies. Developers are working to improve the efficiency and durability of solar panels, seeking to make solar energy even more affordable and efficient. The introduction of new materials and designs could make it possible to increase solar energy production and reduce installation costs.

In addition, smart energy management systems are being developed to optimize the use of energy produced by solar panels. These systems allow real-time monitoring and regulation of energy production and consumption, enabling more efficient management of the power grid.

At the same time, new agricultural technologies are being explored that can be integrated into the Agrivoltaic environment. The use of sensors and crop monitoring systems can provide detailed information on plant needs, enabling more precise irrigation and nutrient management.

The use of precision agriculture techniques, such as the **use of drones to map crops**, can also help farmers optimize production and reduce environmental impact.

Agrivoltaic-related business models are also evolving, with new opportunities for additional income for farmers through energy sales and collaborations between farms and solar energy providers.

Finally, research and development continue to be critical to Agrivoltaics. Studies on long-term energy and agricultural performance, the effect of shading on crops, and energy flow analysis help drive innovation and improve understanding of the impacts and benefits of Agrivoltaics.

Potential global impact of Agrivoltaics on sustainability

Agrivoltaics has the potential to have a significant impact on sustainability globally. This integrated approach, which combines solar energy production with agricultural activity, offers several benefits in terms of clean energy production, reduced carbon emissions, increased resilience of farming communities, conservation of natural resources, and promotion of food security.

The shared use of land for growing food and generating renewable energy reduces pressure on the land and preserves natural resources, contributing to the conservation of local ecosystems. In addition, Agrivoltaics provides additional income opportunities for farmers and promotes local food production, reducing import dependence and promoting long-term sustainability. Widely implementing Agrivoltaics can thus contribute significantly to environmental, energy and food sustainability globally.

CONCLUSIONS

Call to action for the adoption of Agrivoltaics

Agrivoltaics represents a very promising solution to address global challenges related to energy and agriculture. To maximize the benefits of this practice, it is critical to promote and encourage the widespread adoption of Agrivoltaics. Here are some actions that can be taken:

Awareness and information: Educate the public, farmers, government agencies and organizations about the nature and benefits of Agrivoltaics. Communicating the environmental, energy and economic benefits can encourage greater understanding and interest in this practice.

Financial support and incentives: Farmers and investors can be encouraged to adopt Agrivoltaics through subsidized financing programs, grants, or tax incentives. These tools can reduce upfront costs and make Agrivoltaics more accessible and cost-effective.

Collaboration between sectors: It is important to promote collaboration between the agricultural and energy sectors. Farmers, solar energy producers, energy service providers, and government agencies can work together to identify opportunities for Agrivoltaics implementation, share knowledge and resources, and develop sustainable business models.

Development of appropriate policies and regulations: Governments should play a key role in the adoption of

Agrivoltaics through the development of policies and regulations that facilitate the integration of agricultural and solar energy activities. This may include simplifying permitting procedures, adjusting energy tariffs to incentivize renewable energy production, and promoting sustainability standards.

Research and development: Investing in research and development is key to improving agrivoltaic technologies and practices. Research can help optimize solar energy production, identify the most suitable crops, and develop efficient management models. In addition, sharing best practices and research outcomes can foster collective learning and accelerate the adoption of Agrivoltaics.

The adoption of Agrivoltaics requires a collective commitment from farmers, businesses, governments and civil society. Action is needed now to harness the full potential of Agrivoltaics and promote a sustainable future in which clean energy and food production can coexist harmoniously, contributing to the conservation of natural resources and mitigation of climate change.

Sign up for our newsletter to be updated on new releases

gs-publisher.eu

Environment, Satire and Education.

www.ingramcontent.com/pod-product-compliance
Lightning Source LLC
Chambersburg PA
CBHW070823220526
45466CB00002B/746

* 9 7 9 8 3 9 8 1 8 5 7 0 6 *